ТООНЫ ТҮҮХ

THE NUMBER STORY

SMALL BOOK ONE

ENGLISH - MONGOLIAN

Numbers Teach Children
Their Number Names

written and illustrated by

MISS ANNA

Early Reader Edition of *The Number Story 1*
Bronze Medal Winner, 2016 Wishing Shelf Book Award

Copyright © 2018 by Jieeun Woo
Illustrations © Jieeun Woo

Cover by | Lumpy Publishing
Layout by | Lumpy Publishing
Translated by Javkhlan Davaasuren (Д. Жавхлан)
Coloring by Jieeun Woo and Maria Mirabella

All rights reserved. No part of this book may be reproduced or transmitted in any form or by any means whatsoever, including photocopying, recording or by any information storage and retrieval system, without written permission from the publisher and/or author: missanna@missannabooks.com.

Library of Congress Control Number: 2018902040

Names: Miss Anna, author.
Title: Number story : numbers teach children their number names / Miss Anna.
Description: Portland, OR: Lumpy Publishing, 2018.
Identifiers: ISBN 978-1-949320-02-2| LCCN 2018902040
Summary: The pictures and rhymes present stories which introduce numbers 0-10.
Subjects: LCSH Numeration—English—Mongolian--Pictorial works--Juvenile literature. | BISAC JUVENILE NONFICTION /
Languages: English—Mongolian
Classification: LCC QA141.3 .M57 2018 | DDC 513—dc23

Publisher: Lumpy Publishing
Website: www.missannabooks.com
Email: missanna@missannabooks.com

Paperback: ISBN 978-1-949320-02-2
Printed in the U.S.A.   1 3 5 7 9 10 8 6 4 2

Тоонуудын нэрийг сурмаар байна уу?

It is very easy and a lot of fun!

Энэ их амархан бас их хөгжилтэй!

Say-along our little jingle

Бидэнтэй хамт дуулаарай!

starting from Number One!

Бүгдээрээ Нэгийн Тооноос эхэлье!

ONE looks like my one finger.

НЭГ

миний хуруу шиг харагддаг.

ONE!
НЭГ!

2

TWO trails a tail.

ХОЁР

бол сүүлтэй.

A TAIL! СҮҮЛ!

3

THREE has bumps.

ГУРАВ

бол овгортой.

BUMPY! OBГOP!

4

FOUR carries a sail.

ДӨРӨВ далбаатай.

A SAIL!
ДАЛБAA!

5

FIVE   is a racing track.

ТАВ

бол уралдааны зам.

VRooom
BPYYM!
1

6

S I X   curves like a snail.

ЗУРГАА

эмгэн хумс шиг муруй.

A Snail!   ЭМГЭН ХУМС!

7

BE CAREFUL! IT'S SHARP!

Болгоомжтой! Энэ их Хурц шүү!

8

EIGHT   is rollercoaster rails.

НАЙМ

бол галзуу хулганы зам.

Гоё доо!
YIPPEE!

NINE   is a bubble on a stick.

ЕС

бол гуурсан дээрх хөөс.

A BUBBLE!

XΘΘC!

10

TEN   is an eye of a whale.

АРАВ

бол халимны нүд.

HELLO!

САЙН УУ!

And

Тэгээд

0

ZERO   is an empty pail.

ТЭГ

бол хоосон сав.

IT'S EMPTY!
Энэ хоосон байна!

Thank you for playing with us today.

We had a lot of fun too!

Бидэнтэй хамт тоглосонд баярлалаа.

Хөгжилтэй, сайхан байлаа!

We are your Number friends,
Zero to Ten,
Who will be here for you~

Бид чиний Тоон найзууд

Тэгээс Арав.

Бид үргэлж чиний дэргэд байна.

Bye-bye now!
See you again soon!

Түр баяртай!

Удахгүй дахин уулзъя!

The Numbers are *SINGING* too!

To sing-a-long, look for Miss Anna Number Story
at your favorite music store like iTUNES.

MP3

Numbers 0-10
IDENTIFYING
& COUNTING

Numbers 11-20
& Ordinals

first, second, third...

Numbers 0-100
& Place Values

ones, tens, hundreds...

About Clocks
& Telling Time

hours, minutes, seconds...

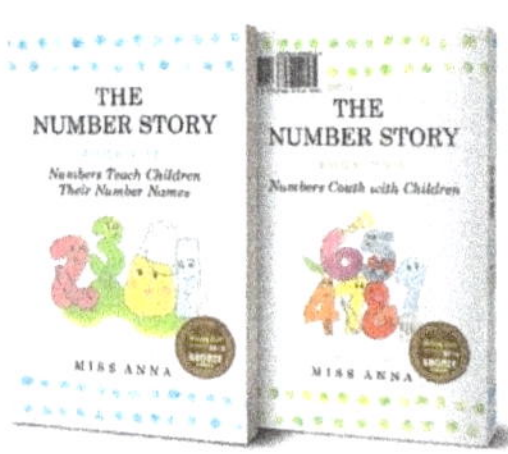

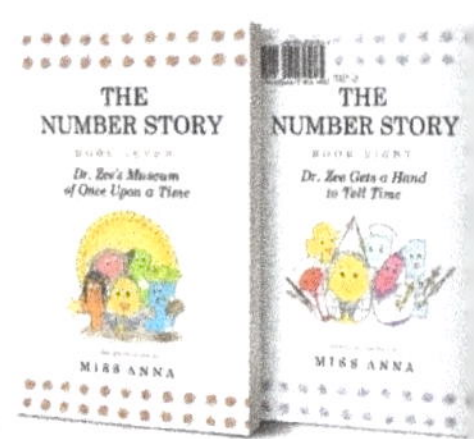

Number Story 1 & 2

isbn: 978-0-996216-48-7

Number Story 3 & 4

isbn: 978-1-945977-01-5

Number Story 5 & 6

isbn: 978-1-945977-06-0

Number Story 7 & 8

isbn: 978-1-949320-40-

For more Miss Anna books to love,
visit us at

www.missannabooks.com

Numbers are working hard all over the world!
Come Travel the World with Us!

www.ingramcontent.com/pod-product-compliance
Lightning Source LLC
Chambersburg PA
CBHW040900070726
47599CB00035B/2251